"Do not worry about your difficulties in Mathematics. I can assure you mine are still greater"

Albert Einstein

Copyright 2017, Sahil Bora

All Rights Reserved

Table of Contents

Introduction 1
My story 2
Go to class 5
Note taking 7
Understanding concepts 11
Completing tutorial questions and problems sets 16
When you don't understand a concept 18
Preparing for an assessment 25
Assessment taking 29
Case study 33
Conclusion 37
About the author 38

Introduction

How do people do well in mathematics?

Is it having natural talent for the subject? Being highly intelligent in dealing with numbers and problems? Studying non-stop till 2 in the morning?

I've heard the phrase "I'm not good at math, because I'm just not a math person" many times around my peers. You know what? During high school I even said it to myself many times. The fact is that this statement is probably the most damaging thing you can say to yourself because you're already putting yourself down.

I'm not saying that everyone can be the next Isaac Newton, Albert Einstein or Terence Tao. It's evident that genetics play a role in having a natural math aptitude. Unfortunately some people are struck with actual learning disabilities like dyslexia or ADHD. I definitely don't want to say that everyone can learn math easily.

However, while there are a few geniuses with natural talent for the subject, using with the phrase, "I'm not a math person" is allowing yourself to believe you're destined for failure and you will never be able to understand the subject. That sir is complete garbage.

An interesting story I read on BetterExplained.com is that "Today, your average 6th grader can understand math with negative numbers, but 300 years ago, it would have been a PhD level area of mathematics research." This would hopefully start to shake the mindset that you need to be a "math person" to understand and do well in the subject.

My story

I was always a struggling maths student especially during my last two years of high school. I knew I wanted to do engineering and computer science after I graduated and the requirements for entry into those courses required me to complete VCE Mathematical Methods. I also wanted to do VCE Specialist Maths in year 11, the highest level of maths you can complete in high school, but after the maths coordinator saw my year 10 maths grades which were pretty average, he forced me to drop out of the subject.

Mathematical Methods in year 12 never went well for me. I pretty much failed every test despite trying my best to learn the material. My teacher hated my guts because he didn't think I was trying at all and thought I wasn't going to pull through.

When the final exams were near, I managed to obtain over 50 practice exams by finding a zip file on the internet containing practice exams from several different tutoring and coaching companies across Victoria, a scare resource which gave me an advantage over my peers.

I was also receiving one on one tutoring outside of school before the final examination period to try to get as much help as I could to score a decent grade. Even from all the tests I failed during the year, I managed to motivate myself to study hard for the final exams, so during the months of September and October before the November examinations I had no social life. My life consisted of just school, after school tutoring, exam study and part time work.

I was rigorously doing all the practice exams I found, trying to get better at each one I was doing. I pretty much did a practice exam every day from September until the day before the first exam.

When the day came for my first Mathematical Methods exam, I felt confident, and energetic with thoughts in my mind that I could ace this exam because I had been preparing for over two months. The first exam was a calculator free exam, so when I opened it and started reading it, I still felt prepared until I read the fourth question. It was a probability question that required me to think outside the box. Later in the paper there were two more probability questions which I could have easily done on a calculator but had to do manually. I was struggling to write solutions, as I couldn't understand each step of solving the problem.

The second exam became more of a nightmare than exam 1, despite being open book and with calculators enabled. I found the long worded questions confusing and the technical jargon added in wasn't helping at all. I was trying to apply concepts I didn't have a deep understanding of into the questions. It was then that I resorted to writing in memorized solutions from previous practice exams to answer the questions. My high hopes of doing well in the final exams of Mathematical Methods all went down the drain.

After I completed VCE in 2012, I managed to get into RMIT University despite my disappointing VCE Results. University was a fresh start for me. I learnt at university that nobody cares if you got an ATAR of 99.95 or below 30, because it's only just a number just to get into a course.

It was then I begin experimenting with ways to learn better. I began taking notes on what worked, what didn't and which methods proved to be most beneficial. Now I want to show you my guidelines and techniques got how I turned from a failing maths student in high school to doing well in mathematics at university.

Go to class

The first rule is always go to class. The importance of this rule cannot be overemphasized. It doesn't matter if the class starts at 8am on a Monday morning. Wake up, get dressed and get to the lecture on time.

If you skip the lecture. It will take twice as long studying to make up for what you missed. You will also need to bother your friends for their notes and learn the information from scratch by yourself. This is why regular attendance is important for mathematic subjects. If you attend class regularly, you will be able to reduce the amount of studying required to score a decent grade. Don't make it negotiable, even if you're tired, hung over or juggling a part time job, find a way to make it there.

Why to attend class regularly

It is found to be common that lecturers often give indicators about what material is worth really knowing and which you don't have to sweat about.

You concentrate better in class. Listening to the lecturer in person, surrounded by your peers allows you to take notes and learn the material quickly instead of trying to do it at home with social media sites Facebook and Instagram distracting you every 5 minutes.

If you skip class more than once, this suddenly becomes an option for the future classes. You start to endure a debate in your mind before every class, and that's a hard battle to win, especially when you're tired or hung over which can occur often. For the sake of doing well in mathematics, you're much better keeping class attendance mandatory and take the option of skipping class off the table.

When it comes to revision time and getting help, if you don't show up to class regularly, you can't really take advantage of getting help. It will be obvious to the professor that you aren't in lectures because you won't know what was said in class and professors do recognize their students.

Don't forget that professors are human beings too. If your borderline between a distinction grade and a credit grade and you're the student that regularly skips class, you can count on getting only a credit. What does it matter to the professor? He/ She doesn't know who you are, so he/she can give you a credit without flinching. However, if the professor knows you from class, it will be much harder for him/ her to not give you the benefit of doubt.

Note taking

The main goal of note taking in learning mathematics

The goal of note taking in learning mathematics is to record as many example problems as possible. Your entire focus while in class is to write down the steady flow of examples provided by your lecturer.

Prioritise your note taking

In a perfect world you would be able to capture every single problem discussed in class with all the steps and solutions. Don't expect this to happen. The lecturers move too quickly for you to record all of the sample problems presented so you must learn how to prioritise your note taking.

Priority 1: Record the sample problem and final solutions

Even in the fastest class, there should be time to jot down the question and the final solutions. If you're in the middle of writing down the steps of the solutions when the lecturer gives the answer and moves on to the next problem, skip the rest of the steps, record the answer and move on too.

You can try to come back and fill in the steps, but even if you don't go back, having only the problem and final solution will still be useful for review later.

Priority 2: Record the steps of the sample problem

Jotting down the steps of the sample problem being presented will allow your mind to understand how the problem is solved and how the concepts taught by the lecturer are being applied to the sample problem. When you have jotted down the steps of the sample problem, this allows you to apply the next step.

Priority 3: Question the confusing

Students who do well in mathematics are those who closely follow the problems being presented and then pose a question when they don't understand a specific step. If you are unable to ask a question, then at least mark where you got confused when you are jotting down the sample problem. If you are able to ask your lecturer, by all means ask. Remember the more questions you get answered in class, the less work you will have to do later filling in the blanks of your understanding.

Priority 4: Annotate the steps

If you get ahead of the lecturer on a sample problem being shown, and you have time to kill, annotate the steps with little explanations on the side of what they accomplish or why it is necessary. When you do have time to jot down these annotations, they will prove to be immensely useful when you review.

Taking note of each concept presented

When you are taking a mathematics course, there will be several concepts spewed out by your lecturer. The best way to keep up to date with all the concepts being presented is to create a list of all the concepts taught in the class.

This will also help when you need to review for a quiz or exam as it breaks down the syllabus into headings. It will allow you to tick off the concepts you have an insight for and the concepts you don't understand.

An example of this is when the lecture is about differentiation. The key is to break down each concept into specific points. For example, with Differentiation, the following list of concepts would be limits, chain rule, product rule, quotient rule and simple derivative rules.

An example of my list of concepts taught in the
Differentiation Calculus module

Concepts in Differentiation

Tutorial 5
- Limits
- Chain rule
- Product rule
- Quotient rule
- Simple derivative rules

Tutorial 6
- Implicit differentiation
- Inverse trignonometric functions differentiation
- Logarithmic differentiation
- Hyperbolic/Inverse Hyperbolic function and their derivatives

Tutorial 7
- Tangent/Normal
- Rates of Change
- Related rate of change

Tutorial 8
- Maximum and minumum problems
- Small Approximation
- Newton's Method

Tutorial 9
- Basic anti differentiation
- Qutionient Fraction
- Partial Fraction
- Integration by part
- Subtitutional Method

Tutorial 10
- Calculating Area
- Volume of revolution
- Trapezoidal rule
- Simpson's 1/3rd Rule

Tutorial 11
- Implicit differentiation
- Double derivative

Understanding concepts

How mathematical classes are taught

Mathematical classes have a simple structure. In each lecture, the lecturer will present a series of concepts. He may cover maybe just one or more than a dozen. For each concept, the professor will provide examples of the concept in practice. That's how a mathematics class runs.

A case study of two students taking a calculus course

Imagine two students, Joe and Tony.

Joe wants to pass his calculus test, so he tries to memorize the formulas and repeatedly does the practice problems.

Tony also wants to pass but is more interested in understanding calculus. He believes it can be interesting by investing time to gather a deep understanding of the concepts being applied in real life problems. He doesn't just look at the formulas and try to memorise them, but instead understand where they originated and how they are used.

The day of the exam comes and a major question worth several marks comes from a word problem. It is asked in a way that neither Joe nor Tony have ever seen before.

Who do you think will get the better grade?

Magic of insight

When you are copying down the sample problems spewed out by your lecturer, you need to do your best to develop insight. What do I mean by developing insight? Forget the equations you copied from the whiteboard, I'm talking about that click in your mind that you understand the concept deep down in your bones.

If you want to do well in mathematics, all you have to do is develop insight for every single concept covered in each lecture. That's how the game of achieving decent grades in mathematics works. There's no shortcut; it's the only way.

Students who skip the insight developing phase of all the concepts being taught are the ones who struggle. They write down all the sample problems shown in class without any idea of the concepts and then study by repeatedly doing the sample problems. Then when they sit down for an assessment, and they are faced with new questions that apply the concepts in a difficult way, they have no idea how to attempt the question. They panic and end up doing poorly.

Without obtaining insight of the concepts, you can't do well.

Let's take an example gaining insight of the derivative in calculus.

The mathematical definition of the derivative is that it measures the sensitivity to change of a quantity determined by another quantity. A formula for calculating derivatives in first principles is given by

$$f'(a) = \lim_{h \to 0} \frac{f(a+h) - f(a)}{h}$$

The graph below represents the function drawn in black and a red tangent line is added to that function. The slope of the red tangent line is equal to the derivate of the point of the function at the red marked point on the graph.

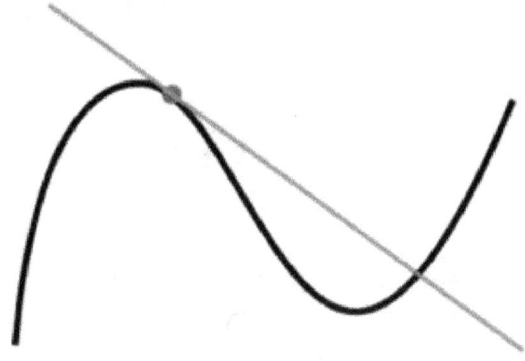

The insight of this complicated definition of the derivative

Forget about the technical mathematical definition you were just given as well as the first principles formula. I'm talking about obtaining developing an insight that you understand deep down.

"The derivate at the red point can be thought as the steepness of the graph at that point". That's it. The complicated definition and formula above is just a way to calculate a specific number that *"describes the steepness"*.

The insight of the derivative can be summarized as *"describing the steepness"*

If you understand the graph, really understand it and are able to teach it someone else simply, you understand the insight behind derivatives.

Here's another example of obtaining an insight of limits.

The technical definition of a limit is the value that a function or sequence approaches as the input or index approaches some value.

$$\lim_{x \to a} f(x) = b$$

Now how can we obtain insight from this technical definition of a limit into something we could explain to a five year old. If we managed to understand the technical definition, an insight that we could develop for this concept could be:

"In the limit, x can be thought of like a stalker, forever getting close to the target, forever trying to get closer to the target but rarely ever succeeding".

The insight of the limit can be summarized as "like a stalker". Later in the book, you will be shown helpful tools to drill down concepts you're trying to understand and obtain magical insight for.

Understanding concepts and gaining insight

"If you can't explain it simply, you don't understand it will enough"
Albert Einstein

The most effective way to see if you understand a concept is to first review it and they try to explain it in your own words as if you were teaching an imaginary class or a five year old. If you can stare at a blank sheet of paper or whiteboard and reproduce a solution easily without a mistake in the process, then you have proven to yourself that you understand the concept.

Passively reviewing a concept by reading and rereading notes is not the same as actively trying to produce it from your own words. You have to make the extra effort to get it into your head and passive review will not really help that.

A great tool to see if you understand the concepts is to ask yourself a "**concept explanation question**".

Examples of a concept explanation question is

-How does the chain rule work?
-What is the purpose of Eigen values and Eigen Vectors?
-How can we find the determinant of a 3x3 matrix without a calculator?

Once you have developed insight for the concept

Once you've developed an insight for the concept being applied, the next thing to do is a small number of practice tutorial questions and problem sets applying it. If you skip trying to understand the concepts, no amount of practice problems will help you get out of an exam disaster. More guidance effectively completing sample problems is covered in the next chapter.

Completing tutorial questions and problems sets

After you have gone to class and taken notes of the sample problems presented by the lecturer, you need to measure your ability in applying those new concepts you learnt in the class. If you don't attempt the tutorial questions and problem sets, you will not have an idea of whether you can solve the questions yourself and be able to apply the concepts on your own. Completing tutorial questions and problem sets should make up 50% of your learning time because it is the most effective way of learning mathematics.

Two main traps that students fall into when they are attempting to complete tutorial question and problem sets are:

Not getting immediate feedback
If you want to learn the material, you need immediate feedback after completing the question with the solutions in hand. Practice without feedback hinders the learning progression of the material.

Doing the questions repeatedly again and again
Students can fall into the trap that you can gain an understanding of the concepts by simply just doing the questions repeatedly again and again. While you can eventually build an understanding through repetition, it's slow and inefficient. Tutorial questions and problem sets

should be used to highlight concepts in which you need to develop a deeper understanding. The best ways to developer a deeper understanding will be discussed in the next chapter.

Deliberate practice

The key to getting better and understanding mathematics is deliberate practice. This doesn't mean doing the question again and again, but challenging yourself with a new question that is beyond your currently ability and seeing if you can answer the question, analyzing your performance after attempting the question and after doing the question, correcting any mistakes you made during the process. Then repeat and repeat again.

When you don't understand a concept

It can be frustrating when you are bombarded with new concepts you have to understand but you lack the insights for all of them. This leads to students using rote memorization to try to gain insight for the concepts, which has been proved to be slow and inefficient. This following chapter will show you how to learn difficult concepts you are struggling to understand without rote memorization.

The steps to understand a concept you don't get at all

1. Get a black piece of paper
2. Write down at the top of the page the concept you want to understand
3. With the help of a textbook or online resources such as Khan Academy, Patrick JMT Tutoring or Better Explained, write down the details of the concept you are trying to learn with the information you are receiving from these resources as if you were teaching it to someone else

The third step, is of course, crucial because you will most likely be repeating some areas of the concept you already understand. However, you will reach a point where you can't explain it simply, and that's the precise gap in your understanding of the concept that you need to fill.

When you have pinpointed what you don't understand, it becomes much easier to find the precise answer.

Example of breaking down the concept of sampling distribution in probability, which could be taught to a five year old

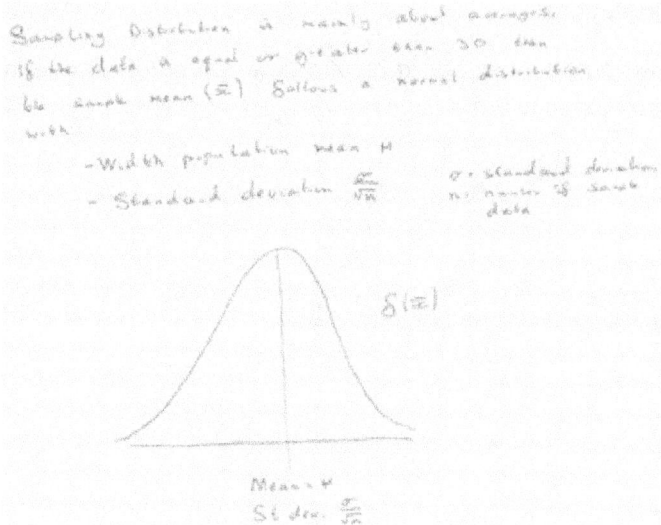

Sampling Distribution is mainly about averages. If the data is equal or greater than 30 then the sample mean ($\bar{x}$) follows a normal distribution with

- With population mean μ
- Standard deviation $\frac{\sigma}{\sqrt{n}}$

σ = standard deviation
n = number of sample data

$\bar{\delta}(\bar{x})$

Mean = μ
St dev. $\frac{\sigma}{\sqrt{n}}$

For understanding the steps of solving a problem

You can use the same steps as you would for trying to understand a concept to fully understand the steps of solving a problem. Break down all the steps and explain not only what it does, but how they execute it with the help of annotations.

An example of solving a second order differential equation step by step with side annotations

Solving 2nd Order Differential Equations

$y'' + 3y' + 2y = 0$ where $y(0)=3$, $y'(0)=5$

① Instead of y substitute in
$$\lambda^2 + 3\lambda + 2 = 0$$

② Find the quadratic roots of the equation
$$\lambda^2 + 3\lambda + 2 = 0 \qquad \text{—— Solve on calculator or algebraically}$$
$$(\lambda + 2)(\lambda + 1) = 0$$
$$\lambda = -1, \lambda = -2$$

③ Since we have two distinct roots we will use the equation
$$y = c_1 e^{-x} + c_2 e^{-2x} \qquad \text{—— Case 1 of 2nd Order Differential equations}$$

④ Solve for $y(0) = 3$
$$3 = c_1 e^{-x} + c_2 e^{-2x}$$
$$3 = c_1 e^0 + c_2 e^0 \qquad \text{—— } e^0 \text{ will equal } 0$$
$$3 = c_1 + c_2$$

⑤ Solve for $y'(0) = 5$
Since $y'(0)=5$, we must differentiate the equation
$$y = c_1 e^{-x} + c_2 e^{-2x}$$
$$y' = -c_1 e^{-x} - 2c_2 e^{-2x}$$

$y'(0) = 5$
$$5 = -c_1 e^{-0} - 2c_2 e^{-0} \qquad \text{—— Substitute } x=0$$
$$5 = -c_1 - 2c_2 \qquad \text{—— Find the value of } c_2$$

$c_1 + c_2 = 3$
$-c_1 - 2c_2 = 5$
$\quad -c_2 = 8$
$\quad c_2 = -8 \qquad \text{—— Now we can find the value of } c_1$
$\quad c_1 = 11$

$$y = 11e^{-x} - 8e^{-2x}$$

Understanding formulas

Formulas should not be memorized, they should be understood. When you see a complicated formula but you can't understand how it works, break down the formula into each part as you would with trying to understand a concept.

An example of trying to break down and understand Euler's Method

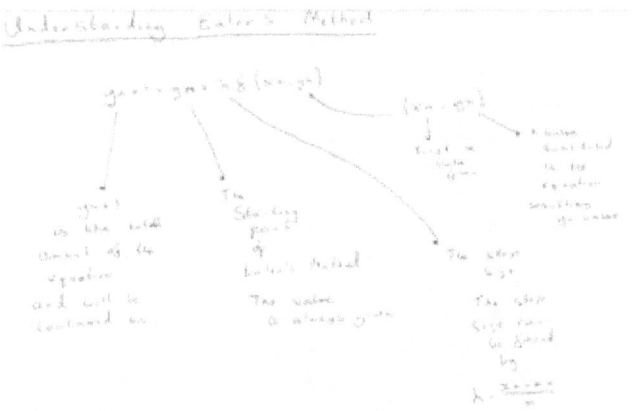

Helpful tools to drill down concepts you're trying to understand

Visualization and Diagramming— Abstract ideas can become more real when they are diagrammed and drawn out. Mathematics has a reputation for being just numbers and analytics, but remember that it's ok to be visual and creative with the concepts you are trying to understand.

An example of visualization being put into practice is diagramming complex numbers. Complex numbers deal with real numbers and imaginary numbers, so how can we visualise this into something we can understand? We can draw out complex numbers graphically on an Argand Diagram to show the relationship with real numbers on the x axis and imaginary numbers on the y axis.

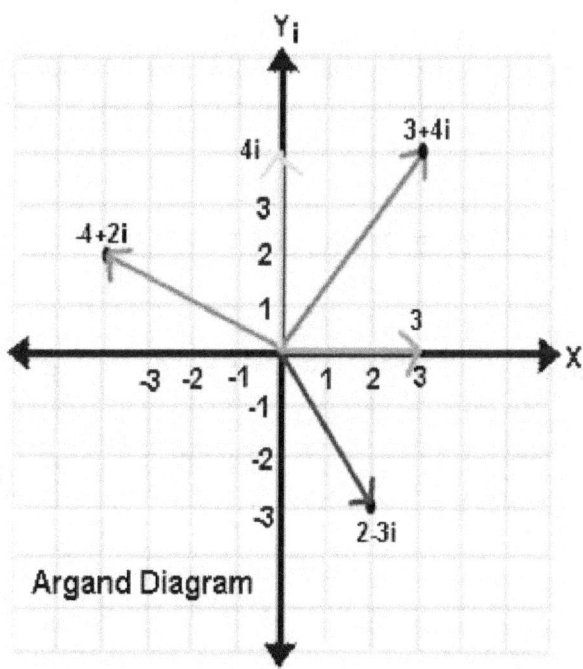

Argand Diagram

Analogies – Relating a concept to an analogy that you understand can work wonders because they are the building blocks for our thoughts. The human brain can easily understand connections between ideas that we already know, hence why they work so well. Analogies being applied were shown in the previous chapter of **Understanding Concepts,** where I tried to break down the two example concepts shown into something understandable.

Simplification – As said earlier in the book, Albert Einstein said, "If you can't explain something simply, then you don't fully understand it." Simplification is a helpful tool for connections between basic and complex ideas.

An example of simplification is trying to understand the determinant in plain, simple English. The technical definition of a determinant is that it is a value associated with a square matrix. To help drill down an understanding, we could say the determinant is just a calculated volume of a matrix.

$$\det\begin{bmatrix} a & b \\ c & d \end{bmatrix} = a \bullet d - c \bullet b$$

$$\det\begin{bmatrix} 8 & 3 \\ 4 & 2 \end{bmatrix} = 8 \bullet 2 - 4 \bullet 3 = 16 - 12 = 4$$

Preparing for an assessment

Students who do well in assessments don't think studying is a big deal. They realize that most of the work required to ace an assessment has already been done by understanding the concepts and solving tutorial/problem sets. By the time the assessment date rolls around, all that's left is a review of the concepts they have mastered and identifying their weak points.

Define the assessment

Before you can conduct any studying, you must define what the assessment will cover. You will need to answer the following questions

Which tutorial/problem sets are most likely to show up?
Is the assessment open book?
Are calculators allowed?
Will the formulas need to memorized?
How much time will be available to complete the assessment?

Cover all the material

A great way to cover all the material is to create an A4 page, with all the concepts from your notes compressed onto one or two pages. Compressing the concepts onto an A4 sheet forces you not only to review all the content, but to simplify it to its basic, important essence.

Constructing Major Problem Sets

Your tutorial/problem sets are the key to your review process. For each lecture relevant to the upcoming assessment, do the following steps:

Step 1: Match the lecture to the sample problems that covers the same material

Step 2: Copy sample problems from the lectures, tutorial questions and problem sets you think are important onto a blank sheet of paper. You don't have to copy the steps or answers, just the question.

Step 3: Always add in concept explanation questions to your major problem sets since they will reveal whether you have gained a deep insight of the concepts or if you've just memorized the definition of the concepts and the steps for the sample problems. Refer to the chapter "**understanding concepts**", which explains concept explanation questions in greater detail.

Step 4: When you start to attempt the major problem sets, begin with the concept explanation questions, thinking about the concepts first and it will make solving the specific problem easier. Try to provide a clear answer for each problem and do your best to give your explanation out loud as if lecturing to class. Otherwise write out the steps and answers clearly. Don't skip any important details.

Step 5: Once you are done with the concept explanation questions, move onto the sample problems. Again, don't do the problems in your head. Write down the important equations and concepts out by hand. Your steps and solutions don't need to be completely perfect as if it they were the actual quiz or exam. But they should clearly demonstrate that you understand what you are doing.

Step 6: If you can't explain exactly how you got from the solution steps to the final answer, then you don't yet understand the problem. Be completely honest with yourself. If you are just regurgitating memorized solutions from previous problems, you aren't prepared to handle new questions on an assessment.

Practice Exams

After covering all the material you are required to know and completing your major problem sets, the most important thing to do is to complete previous practice exams. You want to get comprehensive feedback on your current ability in everything that you will be tested on. Not only are past practice exams comprehensive, but they have a similar format to the new problems you will be facing. Anything other than previous practice exams will be less reliable.

Students who perform poorly on exams also have the problem on estimating what they actually know. One of the hardest problems in successful learning is simply determining what you know and what you don't know. By completing practice exams, you can pinpoint on the points you're struggling on.

Gaining a deeper insight

Once you have identified your weak points from completing practice exams, you need to go back to the previous chapter on "**When you don't understand the concept**" to develop a deeper insight of the concept, and then go back the practice exam you were doing and reattempt the question. If you can reattempt the question without looking at the answers, then you have now understood the weak point you were struggling with.

Working hard and not long

Efficient studying is tiring. Deliberate practice on problems and forcing yourself to gain a deeper insight of concepts is far more draining than skimming lecture notes. These systems force you to learn rapidly and provide feedback to give you confidence. Taking an exam is much easier when you have mastered the insights, completed the tutorial/problem sets with ease and scored well on the practice exams.

Assessment taking

A common problem students face when they are taking assessments is that they run out time to complete the paper. If you try to avoid spending too much time figuring out what the question is asking, then you are likely to provide incomplete answers. On the other hand, if you try to provide very detailed solutions for your answers, then you are likely to run out of time.

With the following tips, you can do your best to eliminate these scenarios and ensure that the grade you receive properly reflects the amount of preparation you put in.

Tip 1: Read all the questions first then answer the questions later

Why should we read all the questions before we answer any? It will allow you to familiarize the length and relative difficulty of what lies ahead in the paper. Your mind will be able to think about all the questions even if you're focusing on trying to understand one difficult question, while another process in your brain will retrieve information about the concepts which you have learnt to be applied to the other questions you are faced with.

This first tip will also help you relax. You either have less than an hour or only a couple of hours to prove what you know and make or break your grade. You begin to question yourself as to whether you are actually prepared to take this assessment. Did you study everything you needed to know? Did you miss any little forgotten concepts you forgot to review?

Just by spending the first couple of minutes carefully reviewing the questions helps break the stress of what students think about when taking an assessment. You begin to say to yourself "All right I think I can actually ace this". Your confidence will begin to arise, your heart rate lowers, and your stress about taking the assessment begins to fade away. You can now turn your full attention to providing excellent solutions to your questions.

Tip 2: Build a time budget with the amount of time you have to complete the assessment

You should know exactly how much time you have to complete the assessment. The key is to lay down very strict time limits for yourself on each question. It keeps you focused and from spending too much time on questions you are finding difficult to answer.

Spending too much time on one question whether it's easy or difficult could jeopardize the chance of finishing the assessment on time. Stay committed to the time limits you have created on each question. Try to finish the assessment 10 minutes early, as this will provide a safety buffer. You want a few minutes available on certain parts to double check your solutions and answers when you are finished, or to go back and be more specific with the solutions on which you rushed.

Tip 3: Proceed from easy to hard

Almost never attempt to answer exam questions in the order they are placed in. The most effective way to tackle an exam is to answer the easiest questions first. Start with the questions you find most approachable before moving onto the more difficult ones. Don't worry if you have skipped all over the assessment; in most cases the other the questions are placed in are irrelevant to each other.

The true advantage of this tip is that it focuses your energy on the questions you know most about which allows you to score maximum points. It also boosts your confidence on tackling more difficult questions that arise. When you tackle hard problems at the end of the assessment, you'll find that the situation seems to be less stressful. You've answered every other question the best you can, so all that's left to do is try to solve the last question you are finding difficult.

Without the pressure of other questions looming in your mind, you can take a more relaxed approach to attempt the puzzler question. You might not have the correct solutions or answers but you can do your best of providing and polishing up a reasonable solution to the question.

Tip 4: Check your work provided you have the time

If you have 10 minutes left that you managed to get with the help of **TIP 2,** always check your work. This will allow you to see if you made any miscalculations in your solutions and if the answers you provided are reasonable. Even putting the mistake of a negative number instead of a positive value can jeopardize the final answers.

If there is a question you didn't feel confident answering, use this time to go over it in detail and make sure you answered it as best you can. It's tempting to proudly walk out early from an exam without checking your answers, but checking your work at the last minute can turn a distinction grade into a high distinction.

Case study

The subject was called Mathematics 1 and it was maths for first year engineering students. The course was equivalent to VCE Specialist maths, with higher level components added in, as well as being taught in a much shorter time frame.

The subject had a high failure rate and as a result of that, if you failed the course, you would have to enroll next semester in the repeat class, which was taught at night till 8pm. I knew I didn't want that to happen, so it did give me a boost of motivation to do well in the subject.

I attended almost every class, as in each lesson we were taught something new and I didn't want to be too far behind in the coursework. It made me concentrate better and learn the material quicker instead of trying to do it myself at home with the internet distracting me.

When the lecturer showed examples of the tutorial problems from the tutorial booklet we received at the start of the semester, I crossed out the ones he showed in class and I attempted the tutorial problems he did not show in class on my own to get immediate feedback on what I did and didn't understand.

When I did come across a concept or problem I didn't understand, I applied the advice of getting a blank sheet of paper with writing down on top of the page the heading of the concept I wanted to understand, and used online resources to drill down and pinpoint what I didn't know. I kept the papers of concepts and problems I was struggling with, as I knew it was going to help me for in the review process for the final exam, which worth 50% of my grade.

When the classes finished for the semester in May, our final exam was going to be in the second week of the exam period, and so I had two weeks to study for the exam. I had plenty of time, so I systematically came up with a plan for how I was going to prepare for the final exam.

Firstly I had to define what the final exam would cover. The final exam wasn't going to cover the first two modules, vectors and complex numbers but it was going to cover the last 4 modules which included differentiation, integration, functions of several variables and power series.

I built a concept list covering all the concepts from those four modules, then created an A4 sheet of paper compressing those concepts as simply as I could for a bird's eye view of all the concepts I either understood well or needed to deepen my intuition.

Creating **major problem sets** was the next thing to do, so I spent a maximum of two hours copying onto a blank sheet of paper important sample problems from the lectures and questions from the tutorial book I was predicting would be on the final, along with adding in **concept explanation questions** to make sure I understood the concept well enough.

When I was attempting my major problem sets I always starting with the conceptual explanation questions, explaining them as if I was teaching an imaginary class thanks to my white board. The actual problems were then attempted, with the solutions written out clearly. If I couldn't explain the concept well or complete the actual problem, I made a note of it so I could later deepen my understanding of the concept.

After I had finished completing my major problem sets and I was happy with gaining a deeper insight of all the things I had trouble with, I started to attempt the previous final exam papers.

I attempted three final exam papers in timed conditions to receive feedback on my current ability to be tested on everything I needed to know. The first exam I received only 65% due to silly mistakes, such as mistyping values on the calculator, and I was still trying to get my head around the topic of power series.

After a drill down of attempting to deepen my understanding of the power series, I attempted my second practice paper with a final result of 85%. As I reviewed the second paper, I found I was struggling with a difficult differentiation question as well as finding a couple of mistakes in calculations of integration problems. My plan to review the difficult differentiation question was to view the answer solutions and break down each step to show the problem was solved.

I attempted my last practice exam two days before the actual exam, I scored 90% which was pleasing but again I was still making small mistakes with a couple of questions which I made note of and reviewed. The day before the actual exam, I just reviewed the practice exams I did with the mistakes I made and I needed to be alert about on the actual exam day and took a good night's rest instead of staying up all night trying to learn everything in one night.

My end grade result for the subject was a very close to a distinction. I performed well on the class quizzes and the final exam but I was just brought down on the mid semester exam as I struggled with complex numbers. From someone who pretty much failed Mathematical Methods in year 12 to very closely achieving a distinction grade in the first year engineering maths, I was completely stoked and proud of my efforts.

Conclusion

Congratulations on finishing reading this book. It doesn't matter if you agree with every single piece of advice you have just encountered; what is important is that by making it this far, you've learnt two valuable points. One this that you can change your grades from failing to exceptional and that there are techniques to help you understand mathematical concepts that work much better.

With this in mind, you are now prepared to transform your grades in mathematics and leap past the majority of your classmates. Good luck on your transformation. You can do it if you take action and put all the advice into practice.

About the author

Sahil Bora is currently a student at RMIT University studying electrical/electronic engineering. How to win at Mathematics is his first book he has written. He is also the author of C++ Better Explained, a book dedicated to teaching the C++ programming language for complete beginners.

In his spare time, Sahil enjoys going surfing and travelling. More information about Sahil Bora can be found at

http://sahilbora.com/

www.ingramcontent.com/pod-product-compliance
Lightning Source LLC
Chambersburg PA
CBHW061232180526
45170CB00003B/1256